NIST Special Publication 800-83
Revision 1

Guide to Malware Incident Prevention and Handling for Desktops and Laptops

Murugiah Souppaya
Computer Security Division
Information Technology Laboratory

Karen Scarfone
Scarfone Cybersecurity
Clifton, VA

July 2013

U.S. Department of Commerce
Cameron F. Kerry, Acting Secretary

National Institute of Standards and Technology
Patrick D. Gallagher, Under Secretary of Commerce for Standards and Technology and Director

Authority

This publication has been developed by NIST to further its statutory responsibilities under the Federal Information Security Management Act (FISMA), Public Law (P.L.) 107-347. NIST is responsible for developing information security standards and guidelines, including minimum requirements for Federal information systems, but such standards and guidelines shall not apply to national security systems without the express approval of appropriate Federal officials exercising policy authority over such systems. This guideline is consistent with the requirements of the Office of Management and Budget (OMB) Circular A-130, Section 8b(3), *Securing Agency Information Systems*, as analyzed in Circular A-130, Appendix IV: *Analysis of Key Sections*. Supplemental information is provided in Circular A-130, Appendix III, *Security of Federal Automated Information Resources*.

Nothing in this publication should be taken to contradict the standards and guidelines made mandatory and binding on Federal agencies by the Secretary of Commerce under statutory authority. Nor should these guidelines be interpreted as altering or superseding the existing authorities of the Secretary of Commerce, Director of the OMB, or any other Federal official. This publication may be used by nongovernmental organizations on a voluntary basis and is not subject to copyright in the United States. Attribution would, however, be appreciated by NIST.

Comments on this publication may be submitted to:

National Institute of Standards and Technology
Attn: Computer Security Division, Information Technology Laboratory
100 Bureau Drive (Mail Stop 8930) Gaithersburg, MD 20899-8930

Reports on Computer Systems Technology

The Information Technology Laboratory (ITL) at the National Institute of Standards and Technology (NIST) promotes the U.S. economy and public welfare by providing technical leadership for the Nation's measurement and standards infrastructure. ITL develops tests, test methods, reference data, proof of concept implementations, and technical analyses to advance the development and productive use of information technology. ITL's responsibilities include the development of management, administrative, technical, and physical standards and guidelines for the cost-effective security and privacy of other than national security-related information in Federal information systems. The Special Publication 800-series reports on ITL's research, guidelines, and outreach efforts in information system security, and its collaborative activities with industry, government, and academic organizations.

Abstract

Malware, also known as malicious code, refers to a program that is covertly inserted into another program with the intent to destroy data, run destructive or intrusive programs, or otherwise compromise the confidentiality, integrity, or availability of the victim's data, applications, or operating system. Malware is the most common external threat to most hosts, causing widespread damage and disruption and necessitating extensive recovery efforts within most organizations. This publication provides recommendations for improving an organization's malware incident prevention measures. It also gives extensive recommendations for enhancing an organization's existing incident response capability so that it is better prepared to handle malware incidents, particularly widespread ones.

Keywords

incident response; information security; malware

Acknowledgments

The authors, Murugiah Souppaya of the National Institute of Standards and Technology (NIST) and Karen Scarfone of Scarfone Cybersecurity, thank their colleagues who assisted with developing this revision of the publication.

Acknowledgments, Original Version of SP 800-83

The authors, Peter Mell of the National Institute of Standards and Technology (NIST) and Karen Kent and Joseph Nusbaum of Booz Allen Hamilton, wish to thank their colleagues who reviewed drafts of this document and contributed to its technical content. The authors would particularly like to acknowledge Tim Grance and Murugiah Souppaya of NIST and Lucinda Gagliano, Thomas Goff, and Pius Uzamere of Booz Allen Hamilton for their keen and insightful assistance throughout the development of the document. The authors would also like to express their thanks to security experts Mike Danseglio (Microsoft), Kurt Dillard (Microsoft), Michael Gerdes (Getronics RedSiren Security Solutions), Peter Szor (Symantec), Miles Tracy (U.S. Federal Reserve System), and Lenny Zeltser (Gemini Systems LLC), as well as representatives from the General Accounting Office, and for their particularly valuable comments and suggestions.

The National Institute of Standards and Technology would also like to express its appreciation and thanks to the Department of Homeland Security for its sponsorship and support of NIST Special Publication 800-83.

Trademark Information

Table of Contents

List of Appendices

List of Figures

Executive Summary

Malware, also known as malicious code, refers to a program that is covertly inserted into another program with the intent to destroy data, run destructive or intrusive programs, or otherwise compromise the confidentiality, integrity, or availability of the victim's data, applications, or operating system. Malware is the most common external threat to most hosts, causing widespread damage and disruption and necessitating extensive recovery efforts within most organizations. Organizations also face similar threats from a few forms of non-malware threats that are often associated with malware. One of these forms that has become commonplace is phishing, which is using deceptive computer-based means to trick individuals into disclosing sensitive information.

This publication provides recommendations for improving an organization's malware incident prevention measures. It also gives extensive recommendations for enhancing an organization's existing incident response capability so that it is better prepared to handle malware incidents, particularly widespread ones. This revision of the publication, Revision 1, updates material throughout the publication to reflect the changes in threats and incidents. Unlike most malware threats several years ago, which tended to be fast-spreading and easy to notice, many of today's malware threats are more stealthy, specifically designed to quietly, slowly spread to other hosts, gathering information over extended periods of time and eventually leading to exfiltration of sensitive data and other negative impacts.

Implementing the following recommendations should facilitate more efficient and effective malware incident response activities for Federal departments and agencies.

Organizations should develop and implement an approach to malware incident prevention.

Organizations should plan and implement an approach to malware incident prevention based on the attack vectors that are most likely to be used currently and in the near future. Because the effectiveness of prevention techniques may vary depending on the environment (i.e., a technique that works well in a managed environment might be ineffective in a non-managed environment), organizations should choose preventive methods that are well-suited to their environment and hosts. An organization's approach to malware incident prevention should incorporate policy considerations, awareness programs for users and information technology (IT) staff, vulnerability and threat mitigation efforts, and defensive architecture considerations.

Organizations should ensure that their policies address prevention of malware incidents.

An organization's policy statements should be used as the basis for additional malware prevention efforts, such as user and IT staff awareness, vulnerability mitigation, threat mitigation, and defensive architecture. If an organization does not state malware prevention considerations clearly in its policies, it is unlikely to perform malware prevention activities consistently and effectively throughout the organization. Malware prevention–related policy should be as general as possible to provide flexibility in policy implementation and to reduce the need for frequent policy updates, but should also be specific enough to make the intent and scope of the policy clear. Malware prevention–related policy should include provisions related to remote workers—both those using hosts controlled by the organization and those using hosts outside of the organization's control (e.g., contractor computers, employees' home computers, business partners' computers, mobile devices).

Organizations should incorporate malware incident prevention and handling into their awareness programs.

Organizations should implement awareness programs that include guidance to users on malware incident prevention. All users should be made aware of the ways that malware enters and infects hosts, the risks that malware poses, the inability of technical controls to prevent all incidents, and the importance of users in preventing incidents, with an emphasis on avoiding social engineering attacks. Awareness programs should also make users aware of policies and procedures that apply to malware incident handling, such as how to identify if a host may be infected, how to report a suspected incident, and what users might need to do to assist with incident handling. In addition, the organization should conduct awareness activities for IT staff involved in malware incident prevention and provide training on specific tasks.

Organizations should have vulnerability mitigation capabilities to help prevent malware incidents.

Organizations should have documented policy, processes, and procedures to mitigate known vulnerabilities that malware might exploit. Because a vulnerability usually can be mitigated through one or more methods, organizations should use an appropriate combination of techniques, including security automation technologies with security configuration checklists and patch management, and additional host hardening measures so that effective techniques are readily available for various types of vulnerabilities.

Organizations should have threat mitigation capabilities to assist in containing malware incidents.

Organizations should perform threat mitigation to detect and stop malware before it can affect its targets. The most commonly used malware threat mitigation technical control is antivirus software; organizations should deploy antivirus software on all hosts for which satisfactory antivirus software is available. Additional technical controls that are helpful for malware threat mitigation include intrusion prevention systems, firewalls, content filtering and inspection, and application whitelisting. The System and Information Integrity family of security controls in NIST Special Publication (SP) 800-53, *Recommended Security Controls for Federal Information Systems and Organizations*, recommends having malware protection mechanisms on various types of hosts, including workstations, servers, mobile computing devices, firewalls, email servers, web servers, and remote access servers.

Organizations should consider using defensive architecture methods to reduce the impact of malware incidents.

No matter how rigorous vulnerability and threat mitigation efforts are, malware incidents will still occur. Organizations should consider altering the defensive architecture of their hosts' software to help mitigate those incidents that still occur. One technique is sandboxing, which is a security model where applications are run within a controlled environment that restricts what operations the applications can perform and isolates them from other applications. Another technique is browser separation, which involves using different web browsers for different types of website access (corporate applications, general access, etc.) Finally, segregation through virtualization techniques separate applications or operating systems from each other through the use of virtualization, such as having one OS instance for corporate applications and another OS instance for all other activity.

Organizations should have a robust incident response process capability that addresses malware incident handling.

As defined in NIST SP 800-61, *Computer Security Incident Handling Guide*, the incident response process has four main phases: preparation, detection and analysis, containment/eradication/recovery, and post-incident activity. Some major recommendations for malware incident handling, by phase or subphase, are as follows:

- **Preparation.** Organizations should perform preparatory measures to ensure that they can respond effectively to malware incidents. Recommended actions include—

  - Building and maintaining malware-related skills within the incident response team

  - Facilitating communication and coordination throughout the organization

  - Acquiring the necessary tools (hardware and software) and resources to assist in malware incident handling

- **Detection and Analysis.** Organizations should strive to detect and validate malware incidents rapidly to minimize the number of infected hosts and the amount of damage the organization sustains. Recommended actions include—

  - Analyzing any suspected malware incident and validating that malware is the cause. This includes identifying characteristics of the malware activity by examining detection sources, such as antivirus software, intrusion prevention systems, and security information and event management (SIEM) technologies.

  - Identifying which hosts are infected by the malware, so that the hosts can undergo the appropriate containment, eradication, and recovery actions. Identifying infected hosts is often complicated by the dynamic nature of malware and computing. Organizations should carefully consider host identification issues before a large-scale malware incident occurs so that they are prepared to use multiple strategies for identifying infected hosts as part of their containment efforts. Organizations should select a sufficiently broad range of identification approaches and should develop procedures and technical capabilities to perform each selected approach effectively when a major malware incident occurs.

  - Prioritizing the handling of each incident based on NIST SP 800-61 guidelines and additional malware-specific criteria

  - Studying the behavior of malware by analyzing it either actively (executing the malware) or forensically (examining an infected host for evidence of malware)

- **Containment.** Malware incident containment has two major components: stopping the spread of malware and preventing further damage to hosts. Nearly every malware incident requires containment actions. In addressing an incident, it is important for an organization to decide which methods of containment to employ initially, early in the response. Organizations should have strategies and procedures in place for making containment-related decisions that reflect the level of risk acceptable to the organization. Containment strategies should support incident handlers in selecting the appropriate combination of containment methods based on the characteristics of a particular situation. Specific containment-related recommendations include the following:

  - It can be helpful to provide users with instructions on how to identify infections and what measures to take if a host is infected; however, organizations should not rely primarily on users for containing malware incidents.

  - If malware cannot be identified and contained by updated antivirus software, organizations should be prepared to use other security tools to contain it. Organizations should also be prepared to submit copies of unknown malware to their security software vendors for analysis, as well as contacting trusted parties such as incident response organizations and antivirus vendors when guidance is needed on handling new threats.

— Organizations should be prepared to shut down or block services used by malware to contain an incident and should understand the consequences of doing so. Organizations should also be prepared to respond to problems caused by other organizations disabling their own services in response to a malware incident.

— Organizations should be prepared to place additional temporary restrictions on network connectivity to contain a malware incident, such as suspending Internet access or physically disconnecting hosts from networks, recognizing the impact that the restrictions might have on organizational functions.

■ **Eradication.** The primary goal of eradication is to remove malware from infected hosts. Because of the potential need for extensive eradication efforts, organizations should be prepared to use various combinations of eradication techniques simultaneously for different situations. Organizations should also consider performing awareness activities that set expectations for eradication and recovery efforts; these activities can be helpful in reducing the stress that major malware incidents can cause.

■ **Recovery.** The two main aspects of recovery from malware incidents are restoring the functionality and data of infected hosts and removing temporary containment measures. Organizations should carefully consider possible worst-case scenarios and determine how recovery should be performed, including rebuilding compromised hosts from scratch or known good backups. Determining when to remove temporary containment measures, such as suspension of services or connectivity, is often a difficult decision during major malware incidents. Incident response teams should strive to keep containment measures in place until the estimated number of infected hosts and hosts vulnerable to infection is sufficiently low that subsequent incidents should be of little consequence. However, even though the incident response team should assess the risks of restoring services or connectivity, management ultimately should be responsible for determining what should be done based on the incident response team's recommendations and management's understanding of the business impact of maintaining the containment measures.

■ **Post-Incident Activity.** Because the handling of malware incidents can be extremely expensive, it is particularly important for organizations to conduct a robust assessment of lessons learned after major malware incidents to prevent similar incidents from occurring. Capturing the lessons learned from the handling of such incidents should help an organization improve its incident handling capability and malware defenses, including identifying needed changes to security policy, software configurations, and malware detection and prevention software deployments.

1. Introduction

1.1 Purpose and Scope

This publication is intended to help a wide variety of organizations understand the threats posed by malware and mitigate the risks associated with malware incidents. In addition to providing background information on the major categories of malware, it provides practical, real-world guidance on preventing malware incidents and responding to malware incidents in an effective, efficient manner. The information presented in this publication is intended to be used as data points entered into a much larger risk management process. See the latest version of NIST SP 800-37, *Guide for Applying the Risk Management Framework to Federal Information Systems* for information on the basics of risk management.[1]

This publication is based on the assumption that the organization already has a general incident response program and capability in place. See the latest version of NIST SP 800-61, *Computer Security Incident Handling Guide* for more information on general incident response.[2] NIST SP 800-61 serves as the foundation for this publication.

1.2 Audience

This document has been created for computer security staff and program managers, technical support staff and managers, computer security incident response teams, and system and network administrators, who are responsible for preventing, preparing for, or responding to malware incidents.

1.3 Document Structure

The remainder of this guide is divided into three major sections. Section 2 defines, discusses, and compares the various categories of malware. Section 3 provides recommendations for preventing malware incidents through several layers of controls. Section 4 explains the malware incident response process, focusing on practical strategies for detection, containment, eradication, and recovery.

The guide also contains several appendices with supporting material. Appendices A and B contain a glossary and an acronym list, respectively. Appendix C lists resources that can help readers gain a better understanding of malware, malware incident prevention, and malware incident handling.

---

[1]   http://csrc.nist.gov/publications/PubsSPs.html#800-37
[2]   http://csrc.nist.gov/publications/PubsSPs.html#800-61-rev2

2.   Understanding Malware Threats

Malware, also known as *malicious code,* refers to a program that is covertly inserted into another program with the intent to destroy data, run destructive or intrusive programs, or otherwise compromise the confidentiality, integrity, or availability of the victim's data, applications, or operating system. Malware is the most common external threat to most hosts, causing widespread damage and disruption and necessitating extensive recovery efforts within most organizations.

This section provides basic information on various forms of malware. It defines common terminology that is used throughout the rest of the document, and it presents fundamental concepts of malware. It does not attempt to explain how these different types of malware work in detail, but rather it highlights the basic characteristics of each type of malware. This section first discusses attacker tools, which are often delivered to targeted hosts via malware, and malware toolkits, which are used by attackers to construct malware. The rest of the section examines forms of malware: traditional, phishing, web-based malware, and advanced persistent threats.

2.1   Forms of Malware

Malware has become the greatest external threat to most hosts, causing damage and requiring extensive recovery efforts within most organizations. The following are the classic categories of malware:

- **Viruses.** A virus self-replicates by inserting copies of itself into host programs or data files. Viruses are often triggered through user interaction, such as opening a file or running a program. Viruses can be divided into the following two subcategories:

  - **Compiled Viruses.** A compiled virus is executed by an operating system. Types of compiled viruses include file infector viruses, which attach themselves to executable programs; boot sector viruses, which infect the master boot records of hard drives or the boot sectors of removable media; and multipartite viruses, which combine the characteristics of file infector and boot sector viruses.

  - **Interpreted Viruses.** Interpreted viruses are executed by an application. Within this subcategory, macro viruses take advantage of the capabilities of applications' macro programming language to infect application documents and document templates, while scripting viruses infect scripts that are understood by scripting languages processed by services on the OS.

- **Worms.** A worm is a self-replicating, self-contained program that usually executes itself without user intervention. Worms are divided into two categories:

  - **Network Service Worms.** A network service worm takes advantage of a vulnerability in a network service to propagate itself and infect other hosts.

  - **Mass Mailing Worms.** A mass mailing worm is similar to an email-borne virus but is self-contained, rather than infecting an existing file.

- **Trojan Horses.** A Trojan horse is a self-contained, nonreplicating program that, while appearing to be benign, actually has a hidden malicious purpose. Trojan horses either replace existing files with malicious versions or add new malicious files to hosts. They often deliver other attacker tools to hosts.

- **Malicious Mobile Code.** Malicious mobile code is software with malicious intent that is transmitted from a remote host to a local host and then executed on the local host, typically without the user's explicit instruction. Popular languages for malicious mobile code include Java, ActiveX, JavaScript, and VBScript.

- **Blended Attacks.** A blended attack uses multiple infection or transmission methods. For example, a blended attack could combine the propagation methods of viruses and worms.

Many, if not most, instances of malware today are blended attacks. Current malware also relies heavily on *social engineering*, which is a general term for attackers trying to trick people into revealing sensitive information or performing certain actions, such as downloading and executing files that appear to be benign but are actually malicious. Because so many instances of malware have a variety of malware characteristics, the classic malware categories listed above (virus, worm, etc.) are considerably less useful than they used to be for malware incident handling. At one time, there were largely different procedures for handling incidents of each malware category; now there is largely one set of procedures for handling all malware incidents, thus nullifying the primary need for having categories.

Another problem with the classic categories is that newer forms of malware do not neatly fit into them. For example, in the growing trend of *web-based malware*, also known as drive-by-download, a user's web browsing is redirected to an infected website, often with little or no use of social engineering techniques. The infected website then attempts to exploit vulnerabilities on the user's host and ultimately to install rootkits or other attacker tools onto the host, thus compromising the host. Although the website is infected, its malware does not infect the user's host; rather, it functions as an attacker tool and installs other attacker tools on the host. Web-based malware is a blended attack of sorts, but its components do not map to the other malware categories.

The classic malware categories do not include *phishing*, which refers to use of deceptive computer-based means to trick individuals into disclosing sensitive personal information.[3] To perform a phishing attack, an attacker creates a website or email that looks as if it is from a well-known organization, such as an online business, credit card company, or financial institution. The fraudulent emails and websites are intended to deceive users into disclosing personal data, usually financial information. For example, phishers might seek usernames and passwords for online banking sites, as well as bank account numbers. Some phishing attacks overlap with web-based malware, because they install keystroke loggers or other attacker tools onto hosts to gather additional personal information.

Organizations should avoid expending substantial time and resources in categorizing each malware incident based on the types of categories expressed above.

2.2 Attacker Tools

Various types of attacker tools might be delivered to a host by malware. These tools allow attackers to have unauthorized access to or use of infected hosts and their data, or to launch additional attacks. Popular types of attacker tools are as follows:

- **Backdoors.** A backdoor is a malicious program that listens for commands on a certain TCP or UDP port. Most backdoors allow an attacker to perform a certain set of actions on a host, such as acquiring passwords or executing arbitrary commands. Types of backdoors include zombies (better known as bots), which are installed on a host to cause it to attack other hosts, and remote

---

[3]   For more information on phishing, including examples of recent phishing attacks, visit the Anti-Phishing Working Group website (http://www.antiphishing.org/). Another good resource is *How Not to Get Hooked by a "Phishing" Scam*, from the Federal Trade Commission (FTC) (http://www.ftc.gov/bcp/edu/pubs/consumer/alerts/alt127.shtm).

administration tools, which are installed on a host to enable a remote attacker to gain access to the host's functions and data as needed.

- **Keystroke Loggers.** A keystroke logger monitors and records keyboard use. Some require the attacker to retrieve the data from the host, whereas other loggers actively transfer the data to another host through email, file transfer, or other means.

- **Rootkits.** A rootkit is a collection of files that is installed on a host to alter its standard functionality in a malicious and stealthy way. A rootkit typically makes many changes to a host to hide the rootkit's existence, making it very difficult to determine that the rootkit is present and to identify what the rootkit has changed.

- **Web Browser Plug-Ins.** A web browser plug-in provides a way for certain types of content to be displayed or executed through a web browser. Malicious web browser plug-ins can monitor all use of a browser.

- **E-Mail Generators.** An email generating program can be used to create and send large quantities of email, such as malware and spam, to other hosts without the user's permission or knowledge.

- **Attacker Toolkits.** Many attackers use toolkits containing several different types of utilities and scripts that can be used to probe and attack hosts, such as packet sniffers, port scanners, vulnerability scanners, password crackers, and attack programs and scripts.

Because attacker tools can be detected by antivirus software, some people think of them as forms of malware. However, attacker tools have no infections capability on their own; they rely on malware or other attack mechanisms to install them onto target hosts. Strictly speaking, attacker tools are not malware, but because they are so closely tied to malware and often detected and removed using the same tools, attacker tools will be covered where appropriate throughout this publication.

2.3 The Nature of Today's Malware

The characteristic of today's malware that most distinguishes it from previous generations of malware is its degree of customization. It has become trivial for attackers to create their own malware by acquiring malware toolkits, such as Zeus, SpyEye, and Poison Ivy, and customizing the malware produced by those toolkits to meet their individual needs. Many of these toolkits are available for purchase, while others are open source, and most have user-friendly interfaces that make it simple for unskilled attackers to create customized, high-capability malware.

Here's an example of what a malware toolkit can do, illustrated by how the resulting attack works.

1. The toolkit sends spam to users, attempting to trick them into visiting a particular website.

2. Users visit the website, which has malicious content provided by the toolkit.

3. The website infects the users' computers with Trojan horses (provided by the toolkit) by exploiting vulnerabilities in the computers' operating systems.

4. The Trojan horses install attacker tools, such as keystroke loggers and rootkits (provided by the toolkit).

Many attackers further customize their malware by tailoring each instance of malware to a particular person or small group of people. For example, many attackers harvest information through social networks, then use that affiliation and relationship information to craft superior social engineering attacks.

Other examples are the frequent use of *spear phishing* attacks, which are targeted phishing attacks, and *whaling* attacks, which are spear phishing attacks targeted at executives and other individuals with access to information of particular interest or value.

Malware customization causes significant problems for malware detection, because it greatly increases the variety of malware that antivirus software and other security controls need to detect and block. When attackers are capable of sending a unique attack to each potential victim, it should not be surprising that largely signature-based security controls, such as antivirus software, cannot keep up with them. Mitigation involves a defense in depth approach, using several different detection techniques to increase the odds that at least one of them can detect the malicious behavior of the customized malware.

In addition to customization, another important characteristic of today's malware is its stealthy nature. Unlike most malware several years ago, which tended to be easy to notice, much of today's malware is specifically designed to quietly, slowly spread to other hosts, gathering information over extended periods of time and eventually leading to exfiltration of sensitive data and other negative impacts. The term *advanced persistent threats* (APTs) is generally used to refer to such types of malware. The attack scenario outlined in the above box could be an example of an advanced persistent threat if it was stealthy. APTs may conduct surveillance for weeks, months, or even years, potentially causing extensive damage to an organization with just one compromise. APTs are also notoriously difficult to remove from hosts, often requiring the host's operating system and applications to be reinstalled and all data restored from known good backups.

In summary, today's malware is often harder to detect, more damaging, and harder to remove than previous generations of malware. And there is no indication that this evolution is at an end. When today's hardest malware problems become routine to address, expect new challenges to emerge.

3. Malware Incident Prevention

This section presents recommendations for preventing malware incidents within an organization. The main elements of prevention are policy, awareness, vulnerability mitigation, threat mitigation, and defensive architecture. Ensuring that policies address malware prevention provides a basis for implementing preventive controls. Establishing and maintaining general malware awareness programs for all users, as well as specific awareness training for the IT staff directly involved in malware prevention–related activities, are critical to reducing the number of incidents that occur through human error. Expending effort on vulnerability mitigation can eliminate some possible attack vectors. Implementing a combination of threat mitigation techniques and tools, such as antivirus software and firewalls, can prevent threats from successfully attacking hosts and networks. Also, using defensive architectures such as sandboxing, browser separation, and segregation through virtualization can reduce the impact of compromises. Sections 3.1 through 3.5 address each of these areas in detail and explain that organizations should implement guidance from each category of recommendations to create an effective layered defense against malware.

When planning an approach to malware prevention, organizations should be mindful of the attack vectors that are most likely to be used currently and in the near future.[4] They should also consider how well-controlled their hosts are (e.g., managed environment, non-managed environment); this has significant bearing on the effectiveness of various preventive approaches. In addition, organizations should incorporate existing capabilities, such as antivirus software deployments and patch management programs, into their malware prevention efforts. However, organizations should be aware that no matter how much effort they put into malware incident prevention, incidents will still occur (e.g., previously unknown types of threats, human error). For this reason, as described in Section 4, organizations should have robust malware incident handling capabilities to limit the damage that malware can cause and restore data and services efficiently.

3.1 Policy

Organizations should ensure that their policies address prevention of malware incidents. These policy statements should be used as the basis for additional malware prevention efforts, such as user and IT staff awareness, vulnerability mitigation, threat mitigation, and defensive architecture (described in Sections 3.2 through 3.5, respectively). If an organization does not state malware prevention considerations clearly in its policies, it is unlikely to perform malware prevention activities consistently and effectively throughout the organization. Malware prevention–related policy should be as general as possible to provide flexibility in policy implementation and to reduce the need for frequent policy updates, but also specific enough to make the intent and scope of the policy clear. Although some organizations have separate malware policies, many malware prevention considerations belong in other policies, such as acceptable use policies, so a separate malware policy might duplicate some of the content of other policies.[5] Malware prevention–related policy should include provisions related to remote workers—both those using hosts controlled by the organization and those using hosts outside of the organization's control (e.g., contractor computers, employees' home computers, business partners' computers, mobile devices).

---

[4] See NIST SP 800-61 for more information on this type of information sharing (e.g., blogs and data feeds from antimalware product vendors, incident response organizations, Information Sharing and Analysis Centers)

[5] For example, many acceptable use policies state that the organization's computing resources should be used only in support of the organization. Personal use of computing resources is a common source of malware incidents; however, because there are several other reasons why an organization might want to restrict personal use of computing resources, this policy consideration is more appropriately addressed in the organization's acceptable use policy than a malware policy.

Common malware prevention–related policy considerations include the following:[6]

■ Requiring the scanning of media from outside of the organization for malware before they can be used

■ Requiring that email file attachments be scanned before they are opened

■ Prohibiting the sending or receipt of certain types of files (e.g., .exe files) via email

■ Restricting or prohibiting the use of unnecessary software, such as user applications that are often used to transfer malware (e.g., personal use of external instant messaging and file sharing services)

■ Restricting the use of removable media (e.g., flash drives), particularly on hosts that are at high risk of infection, such as publicly accessible kiosks

■ Specifying which types of preventive software (e.g., antivirus software, content filtering software) are required for each type of host (e.g., email server, web server, laptop, smart phone) and application (e.g., email client, web browser), and listing the high-level requirements for configuring and maintaining the software (e.g., software update frequency, host scan scope and frequency)

■ Restricting or prohibiting the use of organization-issued and/or personally-owned mobile devices on the organization's networks and for telework/remote access.

3.2 Awareness

An effective awareness program explains proper rules of behavior for use of an organization's IT hosts and information. Accordingly, awareness programs should include guidance to users on malware incident prevention, which can help reduce the frequency and severity of malware incidents. All users should be made aware of the ways in which malware enters and infects hosts; the risks that malware poses; the inability of technical controls to prevent all incidents; and the importance of users in preventing incidents, with an emphasis on avoiding social engineering attacks (as discussed below). In addition, the organization's awareness program should cover the malware incident prevention considerations in the organization's policies and procedures, as described in Section 3.1, as well as generally recommended practices for avoiding malware incidents. Examples of such practices are as follows:

■ Not opening suspicious emails or email attachments, clicking on hyperlinks, etc. from unknown or known senders, or visiting websites that are likely to contain malicious content

■ Not clicking on suspicious web browser popup windows

■ Not opening files with file extensions that are likely to be associated with malware (e.g., .bat, .com, .exe, .pif, .vbs)

■ Not disabling malware security control mechanisms (e.g., antivirus software, content filtering software, reputation software, personal firewall)

■ Not using administrator-level accounts for regular host operation

■ Not downloading or executing applications from untrusted sources.

---

[6] Although all of these considerations are intended to help organizations prevent malware incidents, many of them could also be helpful in detecting or containing incidents.

7

As described in Section 4, organizations should also make users aware of policies and procedures that apply to malware incident handling, such as how to identify if a host may be infected, how to report a suspected incident, and what users might need to do to assist with incident handling (e.g., updating antivirus software, scanning hosts for malware). Users should be made aware of how the organization will communicate notices of major malware incidents and given a way to verify the authenticity of all such notices. In addition, users should be aware of changes that might be temporarily made to the environment to contain an incident, such as disconnecting infected hosts from networks.

As part of awareness activities, organizations should educate their users on the social engineering techniques that are employed to trick users into disclosing information. Examples of recommendations for avoiding phishing attacks and other forms of social engineering include:

- Never reply to email requests for financial or personal information. Instead, contact the person or the organization at the legitimate phone number or website. Do not use the contact information provided in the email, and do not click on any attachments or hyperlinks in the email.

- Do not provide passwords, PINs, or other access codes in response to emails or unsolicited popup windows. Only enter such information into the legitimate website or application.

- Do not open suspicious email file attachments, even if they come from known senders. If an unexpected attachment is received, contact the sender (preferably by a method other than email, such as phone) to confirm that the attachment is legitimate.

- Do not respond to any suspicious or unwanted emails. (Asking to have an email address removed from a malicious party's mailing list confirms the existence and active use of that email address, potentially leading to additional attack attempts.)

Although user awareness programs are increasingly important to help reduce the frequency and severity of social engineering-driven malware incidents, the impact of these programs is still typically not as great as that of the technical controls described in Sections 3.3 through 3.5 for malware incident prevention. An organization should not rely on user awareness as its primary method of preventing malware incidents; instead, the awareness program should supplement the technical controls to provide additional protection against incidents.

The awareness program for users should also serve as the foundation for awareness activities for the IT staff involved in malware incident prevention, such as security, system, and network administrators. All IT staff members should have some basic level of awareness regarding malware prevention, and individuals should be trained in the malware prevention–related tasks that pertain to their areas of responsibility. In addition, on an ongoing basis, some IT staff members (most likely, some members of the security or incident response teams) should receive and review bulletins on types of new malware threats, assess the likely risk to the organization, and inform the necessary IT staff members of the new threat so that incidents can be prevented. IT staff awareness activities related to malware incident handling are discussed in Section 4.

3.3 Vulnerability Mitigation

As described in Section 2, malware often attacks hosts by exploiting vulnerabilities in operating systems, services, and applications. Consequently, mitigating vulnerabilities is very important to the prevention of malware incidents, particularly when malware is released shortly after the announcement of a new vulnerability, or even before a vulnerability is publicly acknowledged. A vulnerability can usually be mitigated by one or more methods, such as applying patches to update the software or reconfiguring the software (e.g., disabling a vulnerable service). Because of the challenges that vulnerability mitigation presents, including handling the continual discovery of new vulnerabilities, organizations should have documented policy, processes, and procedures for vulnerability mitigation.

Organizations should consider using security automation technologies with OS and application configuration checklists to help administrators secure hosts consistently and effectively. Security automation technologies can use checklists to apply configuration settings that improve the default level of security and to continuously monitor the hosts' settings to verify that they are still in compliance with the checklist settings.[7] Organizations should also consider using security automation technologies for OS and application patch management—to identify, acquire, distribute, and install security-related patches so as to mitigate vulnerabilities that the patches address.[8]

In terms of security configurations, organizations should use sound host hardening principles. For example, organizations should follow the principle of least privilege, which refers to configuring hosts to provide only the minimum necessary rights to the appropriate users, processes, and hosts. Least privilege can be helpful in preventing malware incidents, because malware often requires administrator-level privileges to exploit vulnerabilities successfully. If an incident does occur, prior application of least privilege might minimize the amount of damage that the malware can cause. Organizations should also implement other host hardening measures that can further reduce the possibility of malware incidents, such as the following:

- Disabling or removing unneeded services (particularly network services), which are additional vectors that malware can use to spread

- Eliminating unsecured file shares, which are a common way for malware to spread

- Removing or changing default usernames and passwords for OSs and applications, which could be used by malware to gain unauthorized access to hosts

- Disabling automatic execution of binaries and scripts, including AutoRun on Windows hosts

- Changing the default file associations for file types that are most frequently used by malware but not by users (e.g., .pif, .vbs) so that such files are not run automatically if users attempt to open them.

Host hardening should also include applications, such as email clients, web browsers, and word processors, that are frequently targeted by malware. Organizations should disable unneeded features and capabilities from these applications, particularly those that are commonly exploited by malware, to limit the possible attack vectors for malware. One example is the use of macro languages by word processors and spreadsheets; most common applications with macro capabilities offer macro security features that permit macros only from trusted locations or prompt the user to approve or reject each attempt to run a macro, thus reducing the chance of macro-induced malware infection. Another example is preventing software installation within web browsers by configuring browsers to prohibit plug-in installation or to prompt users to approve the installation of each plug-in.

---

[7] For more information, see NIST SP 800-128, *Guide for Security-Focused Configuration Management of Information Systems* (http://csrc nist.gov/publications/PubsSPs html#800-128).

[8] For more information on security automation and checklists, see NIST SP 800-70 Revision 2, *National Checklist Program for IT Products: Guidelines for Checklist Users and Developers* (http://csrc nist.gov/publications/PubsSPs html#800-70) and NIST SP 800-117, *Guide to Adopting and Using the Security Content Automation Protocol (SCAP)* (http://csrc nist.gov/publications/PubsSPs.html#800-117). More information on patch management is available from NIST SP 800-40 Revision 3, *Guide to Enterprise Patch Management Technologies* (http://csrc nist.gov/publications/PubsSPs.html#800-40-rev3).

Being able to alter application configuration settings quickly can be very beneficial in remediating vulnerabilities very quickly, including temporary remediation measures. For example, a configuration change could disable a vulnerable service temporarily while the service's vendor prepares and releases a patch that permanently fixes the vulnerability. Once the patch is available and deployed, the organization can reverse the configuration change to reactivate the no longer vulnerable service. Organizations should consider in advance how configuration settings could be changed in response to a malware emergency and establish and maintain appropriate procedures.

3.4   Threat Mitigation

Organizations should perform threat mitigation to detect and stop malware before it can affect its targets. Even if virtually all vulnerabilities in a host have been mitigated, threat mitigation is still critically important—for example, for stopping instances of malware that do not exploit vulnerabilities, such as attacks that rely on social engineering methods to trick users into running malicious files. Threat mitigation is also critical for situations where a major new threat is likely to attack an organization soon and the organization does not have an acceptable vulnerability mitigation option. For example, there might not be a patch available for a new vulnerability.

This section describes several types of security tools that can mitigate malware threats: antivirus software, intrusion prevention systems (IPS), firewalls, content filtering/inspection, and application whitelisting. For each of these categories, the section also describes typical features, the types of malware and attack vectors the tools address, and the methods they use to detect and stop malware. Recommendations and guidance for implementing, configuring, and maintaining the tools are also provided, as well as explanations of the tools' shortcomings and the ways in which they complement other tools. In addition, the section discusses client and server application settings that can be helpful in mitigating threats.

3.4.1   Antivirus Software

Antivirus software is the most commonly used technical control for malware threat mitigation. There are many brands of antivirus software, with most providing similar protection through the following recommended capabilities:

- ■ Scanning critical host components such as startup files and boot records.

- ■ Watching real-time activities on hosts to check for suspicious activity; a common example is scanning all email attachments for known malware as emails are sent and received. Antivirus software should be configured to perform real-time scans of each file as it is downloaded, opened, or executed, which is known as *on-access scanning*.

- ■ Monitoring the behavior of common applications, such as email clients, web browsers, and instant messaging software. Antivirus software should monitor activity involving the applications most likely to be used to infect hosts or spread malware to other hosts.

- ■ Scanning files for known malware. Antivirus software on hosts should be configured to scan all hard drives regularly to identify any file system infections and, optionally, depending on organization security needs, to scan removable media inserted into the host before allowing its use. Users should also be able to launch a scan manually as needed, which is known as *on-demand scanning*.

- ■ Identifying common types of malware as well as attacker tools.

- *Disinfecting* files, which refers to removing malware from within a file, and *quarantining* files, which means that files containing malware are stored in isolation for future disinfection or examination. Disinfecting a file is generally preferable to quarantining it because the malware is removed and the original file restored; however, many infected files cannot be disinfected. Accordingly, antivirus software should be configured to attempt to disinfect infected files and to either quarantine or delete files that cannot be disinfected.

Organizations should use both host-based and network-based antivirus scanning. Organizations should deploy antivirus software on all hosts for which satisfactory antivirus software is available. Antivirus software should be installed as soon after OS installation as possible and then updated with the latest signatures and antivirus software patches (to eliminate any known vulnerabilities in the antivirus software itself). The antivirus software should then perform a complete scan of the host to identify any potential infections. To support the security of the host, the antivirus software should be configured and maintained properly so that it continues to be effective at detecting and stopping malware. Antivirus software is most effective when its signatures are fully up-to-date. Accordingly, antivirus software should be kept current with the latest signature and software updates to improve malware detection.

Organizations should use centrally managed antivirus software that is controlled and monitored regularly by antivirus administrators, who are also typically responsible for acquiring, testing, approving, and delivering antivirus signature and software updates throughout the organization. Users should not be able to disable or delete antivirus software from their hosts, nor should they be able to alter critical settings. Antivirus administrators should perform continuous monitoring to confirm that hosts are using current antivirus software and that the software is configured properly. Implementing all of these recommendations should strongly support an organization in having a strong and consistent antivirus deployment across the organization.

A possible measure for improving malware prevention is to use multiple antivirus products for key hosts, such as email servers. For example, one antivirus vendor might have a new signature available several hours before another vendor, or an organization might have an operational issue with a particular signature update. Another possibility is that an antivirus product itself might contain an exploitable vulnerability; having an alternative product available in such cases could provide protection until the issue with the primary product has been resolved. Because running multiple antivirus products on a single host simultaneously is likely to cause conflicts between the products, if multiple products are used concurrently, they should be installed on separate hosts. For example, one antivirus product could be used on perimeter email servers and another on internal email servers. This could provide more effective detection of new threats, but also would necessitate increased administration and training, as well as additional hardware and software costs.

Although antivirus software has become a necessity for malware incident prevention, it is not possible for antivirus software to stop all malware incidents. As discussed previously in this section, antivirus software does not excel at stopping previously unknown threats. Antivirus software products detect malware primarily by looking for certain characteristics of known instances of malware. This is highly effective for identifying known malware, but is not so effective at detecting the highly customized, tailored malware increasingly being used.

3.4.2   Intrusion Prevention Systems

Network-based intrusion prevention systems (IPS) perform packet sniffing and analyze network traffic to identify and stop suspicious activity.[9] Network-based IPS products are typically deployed *inline*, which means that the software acts like a network firewall. It receives packets, analyzes them, and allows acceptable packets to pass through. The network-based IPS architecture allows some attacks to be detected on networks before they reach their intended targets. Most network-based IPS products use a combination of attack signatures and analysis of network and application protocols, which means that they compare network activity for frequently attacked applications (e.g., email servers, web servers) to expected behavior to identify potentially malicious activity.

Network-based IPS products are used to detect many types of malicious activity besides malware, and typically can detect only a few instances of malware by default, such as recent major worms. However, some IPS products are highly customizable, allowing administrators to create and deploy attack signatures for many major new malware threats in a matter of minutes. Although there are risks in doing this, such as a poorly written signature triggering false positives that block benign activity inadvertently, a custom signature can block a new malware threat hours before antivirus signatures become available. Network-based IPS products can be effective at stopping specific known threats, such as network service worms, and email–borne malware with easily recognizable characteristics (e.g., subject, attachment filename).

Another form of IPS, known as a *network behavior analysis (NBA) system*, attempts to stop attacks by identifying unusual network traffic flows. Although these products are primarily intended to stop distributed denial of service (DDoS) attacks against an organization, they can also be used to identify worm activity and other forms of malware, as well as use of attacker tools such as backdoors and email generators. An NBA system typically works by monitoring normal network traffic patterns, including which hosts communicate with each other using which protocols, and the typical and peak volumes of activity, to establish baselines. The software then monitors network activity to identify significant deviations from the baselines. If malware causes a particularly high volume of network traffic or uses network or application protocols that are not typically seen, an NBA system should be able to detect and, if deployed inline, block the activity. Another way of limiting some malware incidents is by configuring network devices to limit the maximum amount of bandwidth that can be used by particular hosts or services. Also, some types of network monitoring software can detect and report significant deviations from expected network activity, although this software typically cannot specifically label the activity as malware-related or block it.

Host-based IPS products are similar in principle and purpose to other IPSs, except that a host-based IPS product monitors the characteristics of a single host and the events occurring within that host. Examples of activity that might be monitored by host-based IPSs include network traffic, host logs, running processes, file access and modification, and host and application configuration changes. Host-based IPS products often use a combination of attack signatures and knowledge of expected or typical behavior to identify known and unknown attacks on hosts. For example, host-based IPS products that monitor attempted changes to files can be effective at detecting viruses attempting to infect files and Trojan horses attempting to replace files, as well as the use of attacker tools, such as rootkits, that often are delivered by malware. If a host-based IPS product monitors the host's network traffic, it offers detection capabilities similar to a network-based IPS's.

---

[9]   Intrusion prevention systems are similar to intrusion detection systems (IDS), except that IPSs can attempt to stop malicious activity, whereas IDSs cannot. This section discusses the use of IPSs, not IDSs, for preventing or containing malware incidents. Section 4 describes how both IPS and IDS technologies can be used for malware incident detection.

See NIST SP 800-94, *Guide to Intrusion Detection and Prevention Systems (IDPS)* for more information on IPSs.[10]

3.4.3 Firewalls

A *network firewall* is a device deployed between networks to restrict which types of traffic can pass from one network to another. A *host-based firewall* is a piece of software running on a single host that can restrict incoming and outgoing network activity for that host only. Both types of firewalls can be useful for preventing malware incidents. Organizations should configure their firewalls with *deny by default* rulesets, meaning that the firewalls deny all incoming traffic that is not expressly permitted. With such rulesets in place, malware could not spread using services deemed unnecessary to the organization.[11] Organizations should also restrict outgoing traffic to the degree feasible, with a focus on preventing the use of prohibited services commonly used by malware.

When a major new malware threat targeting a network service is impending, organizations might need to rely on firewalls to prevent an incident. To prepare for worst-case situations, organizations should be ready to add or change firewall rules quickly to prevent a network service–based malware incident. Firewall rules might also be helpful in stopping malware that relies on particular IP addresses, such as a worm that downloads Trojan horses from one of five external hosts. Adding a rule that blocks all activity involving the external hosts' IP addresses could prevent the Trojan horses from reaching the organization.

More information on firewalls is available from NIST SP 800-41 Revision 1, *Guidelines on Firewalls and Firewall Policy* (http://csrc.nist.gov/publications/PubsSPs.html#800-41).

3.4.4 Content Filtering/Inspection

Organizations should use content inspection and filtering technologies for stopping email-based malware threats. Organizations should use spam filtering technologies to reduce the amount of spam that reaches users.[12] Spam is often used for malware delivery, particularly phishing attacks, so reducing spam should lead to a corresponding decline in spam-triggered malware incidents. Organizations should also consider configuring their email servers and clients to block attachments with file extensions that are associated with malicious code (e.g., .pif, .vbs), and suspicious file extension combinations (e.g., .txt.vbs, .htm.exe). However, this might also inadvertently block legitimate activity. Some organizations alter suspicious email attachment file extensions so that a recipient would have to save the attachment and rename it before running it, which can be a good compromise between functionality and security.

Organizations should also use content inspection and filtering technologies for stopping web-based malware threats. Web content filtering software has several ways of doing this; although it is typically thought of as preventing access to materials that are inappropriate for the workplace, it may also contain blacklist and reputation information (see below). It can also block undesired file types, such as by file extension or by mobile code type. For particularly high security situations, organizations should consider

---

[10] http://csrc.nist.gov/publications/PubsSPs.html#800-94

[11] The use of some services cannot be blocked easily through firewall rulesets. For example, some peer-to-peer file sharing services and instant messaging services can use port numbers designated for other services, such as HTTP or Simple Mail Transfer Protocol (SMTP). Attempting to prevent the use of such services by blocking port numbers might cause legitimate services to be blocked. In such cases, it might be necessary to block access to particular IP addresses that host portions of the services, such as instant messaging servers. Also, as described later in this section, application proxies can identify some instances in which one service is used when another is expected.

[12] In addition to the standard spam filtering technologies, there are also emerging solutions that aim to reduce spam through email authentication. An example is the Domain-based Message Authentication, Reporting & Conformance (DMARC) specification (http://www.dmarc.org/).

restricting which types of mobile code (such as unsigned ActiveX) may or may not be used from various sources (e.g., internal servers, external servers).

Organizations should also block undesired web browser popup windows, as a form of content filtering. Some popup windows are crafted to look like legitimate system message boxes or websites, and can trick users into going to phony websites, including sites used for phishing, or authorizing changes to their hosts, among other malicious actions. Most web browsers can block popup windows, and third-party popup blockers are also available.

Both email and web content filtering should use real-time blacklists, reputation services, and other similar mechanisms whenever feasible to avoid accepting content from known or likely malicious hosts and domains. These mechanisms use a variety of techniques to identify certain IP addresses, domain names, or URIs as being probably malicious or probably benign. Real-time blacklists tend to be based on observed malware activity, while reputation services may be based on user opinions or on automated analysis of websites, emails, etc. without necessarily detecting malware. Because the fidelity and accuracy of these mechanisms varies widely from one implementation to another, organizations should carefully evaluate any real-time blacklists, reputation services, or other similar mechanisms before deploying them into production environments to minimize disruption to operations.

3.4.5 Application Whitelisting

Application whitelisting technologies, also known as application control programs, are used to specify which applications are authorized for use on a host. Most application whitelisting technologies can be run in two modes: audit and enforcement. In enforcement mode, the technology generally prohibits all applications that are not in the whitelist from being executed. In audit mode, the technology logs all instances of non-whitelisted applications being run on the host, but does not act to stop them. The tradeoff between enforcement mode and audit mode is simple; using enforcement mode will stop malware from executing, but it may also prevent benign applications not included on the whitelist from being run. Organizations deploying application whitelisting technologies should consider first deploying them in audit mode, so as to identify any necessary applications missing from the whitelist, before reconfiguring them for enforcement mode. Running application whitelisting technologies in audit mode is analogous to intrusion detection software without intrusion prevention capabilities; it can be useful after an infection occurs to determine which hosts were affected, but it has no ability to prevent infections.

Organizations with high security needs or high-risk environments should consider the use of application whitelisting technologies for their managed hosts. Application whitelisting technologies are built into many operating systems and are also available through third-party utilities.

3.5 Defensive Architecture

No matter how rigorous vulnerability and threat mitigation efforts are, malware incidents will still occur. This section describes four types of complementary methods that organizations should consider using to alter the defensive architecture of a host's software so as to reduce the impact of incidents: BIOS protection, sandboxing, browser separation, and segregation through virtualization.

3.5.1 BIOS Protection

Unauthorized modification of BIOS firmware by malicious software constitutes a significant threat because of the BIOS's unique and privileged position within the PC architecture. A malicious BIOS modification could be part of a sophisticated, targeted attack on an organization—either a permanent denial of service (if the BIOS is corrupted) or a persistent malware presence (if the BIOS is implanted with malware). The move from conventional BIOS implementations to implementations based on the Unified Extensible Firmware Interface (UEFI) may make it easier for malware to target the BIOS in a widespread fashion, as these BIOS implementations are based on a common specification.[13] NIST Special Publication 800-147[14] provides guidelines for significantly improving BIOS protection and integrity, which is a necessary foundation for the other host-based methods.

3.5.2 Sandboxing

Sandboxing refers to a security model where applications are run within a sandbox—a controlled environment that restricts what operations the applications can perform and that isolates them from other applications running on the same host. In a sandbox security model, typically only authorized "safe" operations may be performed within the sandbox; the sandbox prohibits applications within the sandbox from performing any other operations. The sandbox also restricts access to system resources, such as memory and the file system, to keep the sandbox's applications isolated from the host's other applications.

Sandboxing provides several benefits in terms of malware incident prevention and handling. By limiting the operations available, it can prevent malware from performing some or all of the malicious actions it is attempting to execute; this could prevent the malware from succeeding or reduce the damage it causes. And the sandboxing environment—the isolation—can further reduce the impact of the malware by restricting what information and functions the malware can access. Another benefit of sandboxing is that the sandbox itself can be reset to a known good state every time it is initialized.

3.5.3 Browser Separation

Multiple brands of Web browsers (e.g., Microsoft Internet Explorer, Mozilla Firefox, Apple Safari, Google Chrome, Opera) can be installed on a single host. Accessing Web sites containing malicious content is one of the most common ways for hosts to be attacked, such as malicious plug-ins being installed within a browser. To reduce the impact of such attacks, users can use one brand of browser for corporate applications and another brand of browser for all other website access. This separates the sensitive corporate data within one browser from the data within the other browser, providing better protection for the corporate data (although this alone cannot adequately secure browser data) and reducing the likelihood that malware encountered during general web browsing will affect corporate applications. Having a separate brand of browser for corporate applications also allows that browser to be secured more tightly, such as disabling all forms of mobile code (e.g., Java, ActiveX) that are not required for the specified applications.

3.5.4 Segregation Through Virtualization

Browser separation essentially segregates web browsers from each other. Virtualization[15] can be used to segregate applications or operating systems from each other, with much more rigor than simple browser separation can provide. For example, an organization could have one OS instance for corporate applications and another OS instance for all other activities, including web browsing. Each OS instance is a known-good virtualized image that contains the appropriate applications and is secured accordingly. The user loads these virtualized images and does their work within these guest OS images, not directly on the host OS itself. A compromise occurring within one image will not affect the other image unless the

---

[13] This paragraph was taken from the Executive Summary of NIST SP 800-147, *BIOS Protection Guidelines*.
[14] http://csrc.nist.gov/publications/PubsSPs.html#800-147
[15] For more information on virtualization, see NIST SP 800-125, *Guide to Security for Full Virtualization Technologies* (http://csrc.nist.gov/publications/PubsSPs.html#800-125).

compromise involves the virtualization software itself. Another benefit is that every time an image is restarted, it can be reloaded from the known-good image, ensuring that any compromises occurring within the image are eradicated.

An alternative strategy, more usable but less secure, is to use a guest OS for more risky behavior (such as general web browsing) and the host OS for corporate applications. This helps to isolate the riskier activities from the other activities on the host. The host OS can be restricted to only whitelisted applications (see Section 3.4.5) to prevent unauthorized applications from being run within it.

4.   Malware Incident Response

As defined in NIST SP 800-61, *Computer Security Incident Handling Guide,* the incident response process has four major phases: preparation, detection and analysis, containment/eradication/recovery, and post-incident activity. Figure 4-1 displays this incident response life cycle. This section of the guide builds on the concepts of SP 800-61 by providing additional details about responding to malware incidents.[16]

The initial phase of malware incident response involves performing preparatory activities, such as developing malware-specific incident handling procedures and training programs for incident response teams. As described in Section 3, the preparation phase also involves using policy, awareness activities, vulnerability mitigation, and security tools to reduce the number of malware incidents. Despite these measures, residual risk will inevitably persist, and no solution is foolproof. Detection of malware infections is thus necessary to alert the organization whenever incidents occur. Early detection is particularly important for malware incidents because they are more likely than other types of incidents to increase their impact over time, so faster detection and handling can help reduce the number of infected hosts and the damage done.

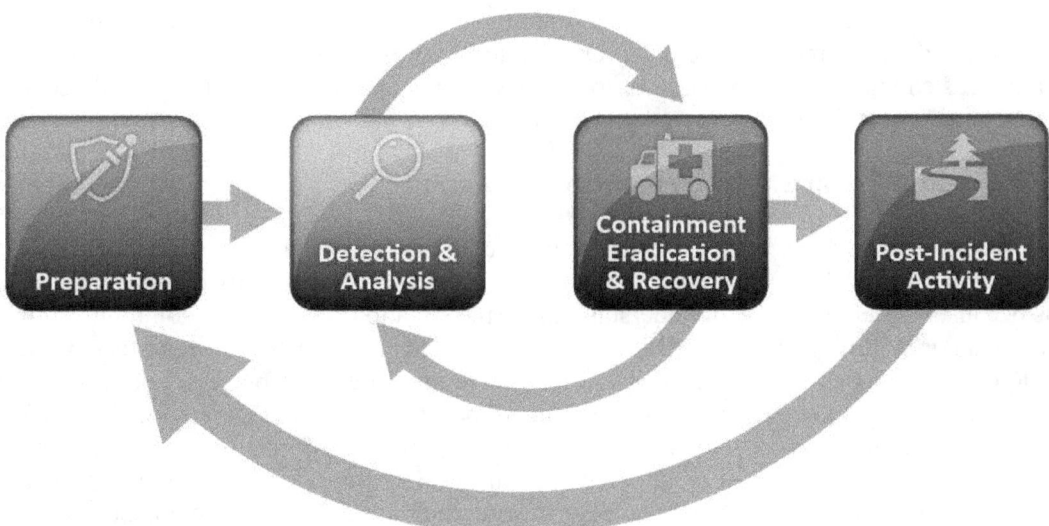

Figure 4-1. Incident Response Life Cycle

For each incident, the organization should act appropriately, based on the severity of the incident, to mitigate its impact by containing it, eradicating infections, and ultimately recovering from the incident. The organization may need to jump back to the detection and analysis phase during containment, eradication, and recovery—for example, to check for additional infections that have occurred since the original detection was done. After an incident has been handled, the organization should issue a report that details the cause and cost of the incident and the steps the organization should take to prevent future incidents and to prepare more effectively to handle incidents that do occur.

This section of the document focuses on those aspects of incident handling that are specific to malware incidents.

---

[16]   For more information on how to establish an incident response capability, refer to NIST SP 800-61, *Computer Security Incident Handling Guide*, available at http://csrc.nist.gov/publications/PubsSPs.html#800-61.

4.1 Preparation

Organizations should perform preparatory measures to ensure that they are capable of responding effectively to malware incidents. Sections 4.1.1 through 4.1.3 describe several recommended preparatory measures, including building and maintaining malware-related skills within the incident response team, facilitating communication and coordination throughout the organization, and acquiring necessary tools and resources.

4.1.1 Building and Maintaining Malware-Related Skills

In addition to standard incident response team skills as described in NIST SP 800-61, all malware incident handlers should have a solid understanding of how each major category of malware infects hosts and spreads. Also, incident handlers should be familiar with the organization's implementations and configurations of malware detection tools so that they are better able to analyze supporting data and identify the characteristics of threats. Incident handlers doing in-depth malware analysis should have strong skills in that area and be familiar with the numerous tools for malware analysis, as described in Section 4.2.4.

Malware incident handlers should keep abreast of the ever-evolving landscape of malware threats and technology. Besides conducting malware-related training and exercises, organizations should also seek other ways of building and maintaining skills. One possibility is to have incident handlers temporarily work as antivirus engineers or administrators so that they can gain new technical skills and become more familiar with antivirus staff procedures and practices.

4.1.2 Facilitating Communication and Coordination

One of the most common problems during malware incident handling is poor communication and coordination. Anyone involved in an incident, including users, can inadvertently cause additional problems because of a limited view or understanding of the situation. To improve communication and coordination, an organization should designate in advance a few individuals or a small team to be responsible for coordinating the organization's responses to malware incidents. The coordinator's primary goal is to maintain situational awareness by gathering all pertinent information, making decisions that are in the best interests of the organization, and communicating pertinent information and decisions to all relevant parties in a timely manner. For malware incidents, the relevant parties often include end users, who might be given instructions on how to avoid infecting their hosts, how to recognize the signs of an infection, and what to do if a host appears to be infected. The coordinator also needs to provide technical guidance and instructions to all staff assisting with containment, eradication, and recovery efforts, as well as giving management regular updates on the status of the response and the current and likely future impact of the incident. Another possible role for the coordinator is interacting with external parties, such as other incident response teams facing similar malware issues.

Organizations should also establish a point of contact for answering questions about the legitimacy of malware alerts. Many organizations use the IT help desk as the initial point of contact and give help desk agents access to sources of information on real malware threats and virus hoaxes so that they can quickly determine the legitimacy of an alert and provide users with guidance on what to do. Organizations should caution users not to forward malware alerts to others without first confirming that the alerts are legitimate.

4.1.3 Acquiring Tools and Resources

Organizations should also ensure that they have the necessary tools (hardware and software) and resources to assist in malware incident handling. See Section 4.2.4 for more information on malware analysis toolkits, systems, and other related resources.

4.2 Detection and Analysis

Organizations should strive to detect and validate malware incidents rapidly to minimize the number of infected hosts and the amount of damage the organization sustains. Because malware can take many forms and be distributed through many means, there are many possible signs of a malware incident and many locations within an organization where the signs might be recorded or observed. It sometimes takes considerable analysis, requiring extensive technical knowledge and experience, to confirm that an incident has been caused by malware, particularly if the malware threat is new and unknown. After malware incident detection and validation, incident handlers should determine the type, extent, and magnitude of the problem as quickly as possible so that the response to the incident can be given the appropriate priority. Sections 4.2.1 through 4.2.4 provide guidance on identifying the characteristics of incidents, identifying infected hosts, prioritizing incident response efforts, and analyzing malware, respectively.

4.2.1 Identifying Malware Incident Characteristics

Because no indicator is completely reliable—even antivirus software might miscategorize benign activity as malicious—incident handlers need to analyze any suspected malware incident and validate that malware is the cause. In some cases, such as a massive, organization-wide infection, validation may be unnecessary because the nature of the incident is obvious. The goal is for incident handlers to be as certain as feasible that an incident is caused by malware and to have a basic understanding of the type of malware threat responsible, such as a worm or a Trojan horse. If the source of the incident cannot easily be confirmed, it is often better to respond as if it were caused by malware and to alter response efforts if it is later determined that malware is not involved. Waiting for conclusive evidence of malware might have a serious negative impact on response efforts and significantly increase the damage sustained by the organization.

As part of the analysis and validation process, incident handlers typically identify characteristics of the malware activity by examining detection sources. Understanding the activity's characteristics is very helpful in assigning an appropriate priority to the incident response efforts and planning effective containment, eradication, and recovery activities. Incident handlers should collaborate with security administrators in advance to identify data sources that can aid in detecting malware information and to understand what types of information each data source may record. In addition to the obvious sources of data, such as antivirus software, intrusion detection system (IDS), and security information and event management (SIEM) technologies, incident handlers should be aware of and use secondary sources as appropriate. See Section 4.2 for more information on possible sources of malware characteristic information.

Once incident handlers have reviewed detection source data and identified characteristics of the malware, the handlers could search for those characteristics in antivirus vendors' malware databases and identify which instance of malware is the most likely cause. If the malware has been known for some time, it is likely that antivirus vendors will have a substantial amount of information on it, such as the following:

- Malware category (e.g., virus, worm, Trojan horse)

- Services, ports, protocols, etc. that are attacked

- Vulnerabilities that are exploited (e.g., software flaws, misconfigurations, social engineering)

- Malicious filenames, sizes, content, and other metadata (e.g., email subjects, web URLs)

- Which versions of operating systems, devices, applications, etc., may be affected

- How the malware affects the infected host, including the names and locations of affected files, altered configuration settings, installed backdoor ports, etc.

- How the malware propagates and how to approach containment

- How to remove the malware from the host.

Unfortunately, the newest threats might not be included in malware databases for several hours or days, depending on the relative importance of the threat, and highly customized threats might not be included in malware databases at all. Therefore, incident handlers may need to consult other sources of information. One option is using public security mailing lists, which might contain first-hand accounts of malware incidents; however, such reports are often incomplete or inaccurate, so incident handlers should validate any information obtained from these sources. Another potentially valuable source of malware characteristic information is peers at other organizations. Other organizations may have already been affected and gathered data on the threat. Establishing and maintaining good relationships with peers at other organizations that face similar problems can be advantageous for all involved. An alternative source of information is self-discovery by performing malware analysis (see Section 4.2.4). This is particularly important if the malware is highly customized; there may be no other way of getting details for the malware other than doing a hands-on analysis.

4.2.2 Identifying Infected Hosts

Identifying hosts that are infected by malware is part of every malware incident. Once identified, infected hosts can undergo the appropriate containment, eradication, and recovery actions. Unfortunately, identifying all infected hosts is often complicated by the dynamic nature of computing. For instance, people shut hosts down, disconnect them from networks, or move them from place to place, making it extremely difficult to identify which hosts are currently infected. In addition, some hosts can boot to multiple OSs or use virtual operating system software; an infection in one OS instantiation might not be detectable when a host is currently using another OS.

Accurate identification of infected hosts can also be complicated by other factors. For example, hosts with unmitigated vulnerabilities might be disinfected and reinfected multiple times. Some instances of malware actually remove some or all traces of other malware, which could cause the partially or fully removed infections to go undetected. In addition, the data concerning infected hosts might come from several sources—antivirus software, IDSs, SIEMs, user reports, and other methods—and be very difficult to consolidate and keep current.

Given the number of malware threats, all infection identification should be performed through automated means (as described in Sections 4.2.2.1 and 4.2.2.2). Manual identification methods, such as relying on users to identify and report infected hosts, and having technical staff personally check each host, are not feasible for most situations. Organizations should carefully consider host identification issues so that they are prepared to use multiple identification strategies as part of implementing effective containment strategies. Organizations should also determine which types of identifying information might be needed and what data sources might record the information. For example, a host's current IP address is typically needed for remote actions; of course, a host's physical location is needed for local actions. One piece of information can often be used to determine others, such as mapping an IP address to a media access control (MAC) address, which could then be mapped to a switch serving a particular group of offices. If an IP address can be mapped to a host owner or user—for example, by recording the mapping during network login—the owner or user can be contacted to provide the host's location.

The difficulty in identifying the physical location of an infected host depends on several factors. In a managed environment, identifying a host's location is often relatively easy because of the standardized manner in which things are done. For example, host names might contain the user's ID or office number, or the host's serial number (which can be tied to a user ID). Also, asset inventory management tools might contain current information on host characteristics. In other environments, especially those in which users have full control over their hosts and network management is not centralized, it might be challenging to link a machine to a location. For example, an administrator might know that the host at address 10.3.1.70 appears to be infected but not have any idea where that machine resides or who uses it. Administrators might need to track down an infected host through network devices. For example, a switch port mapper can poll switches for a particular IP address and identify the switch port number and host name associated with that IP address. If the infected host is several switches away, it can take hours to track down a single machine; if the infected host is not directly switched, the administrator might still need to manually trace connectivity through various wiring closets and network devices. An alternative is to pull the network cable or shut down the switch port for an apparently infected host and wait for a user to report an outage. This approach can inadvertently cause a loss of connectivity for small numbers of uninfected hosts, but if performed carefully as a last-resort identification and containment method, it can be quite effective.

Some organizations first make reasonable efforts to identify infected hosts and perform containment, eradication, and recovery efforts on them, then implement measures to prevent hosts that have not been verified as uninfected and properly secured from attaching to the network. These measures should be discussed well in advance, and incident handlers should have prior written permission to lock out hosts under certain circumstances. Generally, lockout measures are based on the characteristics of particular hosts, such as MAC addresses or static IP addresses, but lockouts can also be performed based on user ID if a host is associated with a single user. Another possibility is to use network login scripts to identify and deny access to infected hosts, but this might be ineffective if an infected host starts spreading malware after system boot but before user authentication. As described in Section 4.3.4, having a separate VLAN for infected or unverified hosts can provide a good way to lock out hosts, as long as the mechanism to detect infections is reliable. Although lockout methods might be needed only under extreme circumstances, organizations should think in advance about how individual hosts or users could be locked out so that if needed, lockouts can be performed rapidly.

Sections 4.2.2.1 through 4.2.2.3 discuss the possible categories of infected host identification techniques: forensic, active, and manual.

4.2.2.1 Forensic Identification

Forensic identification is the practice of identifying infected hosts by looking for evidence of recent infections. The evidence may be very recent (only a few minutes old) or not so recent (hours or days old); the older the information is, the less accurate it is likely to be. The most obvious sources of evidence are those that are designed to identify malware activity, such as antivirus software, content filtering (e.g., anti-spam measures), IPS, and SIEM technologies. The logs of security applications might contain detailed records of suspicious activity, and might also indicate whether a security compromise occurred or was prevented.

In situations in which the typical sources of evidence do not contain the necessary information, organizations might need to turn to secondary sources, such as the following:

■ **DNS Server Logs.** DNS server logs often contain records of infected hosts attempting to get the IP address for an external malicious site that they are trying to interact with (e.g., send data to, receive commands from). Some organizations deploy passive DNS collection systems, which keep track of all DNS resolutions occurring within the enterprise; these are often more helpful than DNS server logs in identifying malicious activity because malware might use DNS services other than the organization's. Analysts should be cautious of blocking hosts based only on resolved IP addresses because many current attacks use fast flux DNS, which means that each domain resolves to several different IP addresses (in a round robin arrangement), and these addresses often change in a matter of hours. The newer the DNS resolution, the more likely the IP addresses are to be the correct ones to block in the short term.

■ **Other Application Server Logs.** Applications commonly used as malware transmission mechanisms, such as email and HTTP, might record information in their logs that indicates which hosts were infected. From end to end, information regarding a single email message might be recorded in several places: the sender's host, each email server that handles the message, and the recipient's host, as well as antivirus and content filtering servers. Similarly, hosts running web browsers can provide a rich resource for information on malicious web activity, including a history of websites visited and the dates and times that they were visited, and cached web data files.

■ **Network Forensic Tools.** Software programs that capture and record packets, such as network forensic analysis tools and packet sniffers, might have highly detailed information on malware activity. However, because these tools record so much information about most or all network activity, it can be very time-intensive to extract just the needed information. More efficient means of identifying infected hosts are often available.

■ **Network Device Logs.** Firewalls, routers, and other filtering devices that record connection activity, as well as network monitoring tools, might be helpful in identifying network connection activity (e.g., specific port number combinations, unusual protocols) consistent with certain malware.

Using forensic data for identifying infected hosts can be advantageous over other methods because the data has already been collected—the pertinent data just needs to be extracted from the total data set. Unfortunately, for some data sources, extracting the data can take a considerable amount of time. Also, event information can become outdated quickly, causing uninfected hosts to undergo containment unnecessarily and allowing infected hosts to avoid containment measures. If an accurate, comprehensive, and reasonably current source of forensic data is available, it might provide the most effective way of identifying infected hosts.

4.2.2.2 Active Identification

Active identification methods are used to identify which hosts are currently infected. Immediately after identifying an infection, some active approaches can be used to perform containment and eradication measures for the host, such as running a disinfection utility, deploying patches or antivirus updates, or moving the host to a VLAN for infected hosts. Active identification can be performed through several methods, including the following:

■ **Security Automation.** Security automation technologies, particularly those used for continuous monitoring (e.g., network access control technologies), can be used to check host characteristics for signs of a current infection, such as a particular configuration setting or a system file with a certain size that indicates an infection. Security automation technologies are generally the preferred method for active identification.

■ **Custom Network-Based IPS or IDS Signature.** Writing a custom IPS or IDS signature that identifies infected hosts is often a highly effective technique. Some organizations have separate IPS or IDS sensors with strong signature-writing capabilities that can be dedicated to identifying malware infections. This provides a high-quality source of information while keeping other sensors from becoming overloaded with malware alerts.

■ **Packet Sniffers and Protocol Analyzers.** Configuring packet sniffers and protocol analyzers to look only for network traffic matching the characteristics of a particular malware threat can be effective at identifying infected hosts. An example of what to monitor is to look for botnet command and control communications being carried over IRC. These packet examination techniques are most helpful if most or all malware-generated network traffic attempts to pass through the same network device or a few devices.

Although active approaches can produce highly accurate results, active approaches need to be used repeatedly because the status of infections changes constantly and the data is gathered over a period of time.

4.2.2.3 Manual Identification

Another method for identifying infected hosts is the manual approach. This is by far the most labor-intensive of the three methods. It should only be considered in those situations where automated methods are not feasible, such as when networks are completely overwhelmed by infection-related traffic using spoofed addresses. Also, if users have full control over their hosts, as they do in many non-managed environments, the characteristics of hosts may be so different that the results of automated identification methods are quite incomplete and inaccurate. In such situations, a manual approach might be needed to supplement automated approaches.

There are a few possible techniques for implementing a manual approach. One is to ask users to identify infections themselves by providing them with information on the malware and the signs of an infection, as well as antivirus software, OS or application patches, or scanning tools. These items may need to be distributed on removable media. A similar manual technique is to have local IT staffers (including individuals who normally do not participate in malware incident handling) either check all hosts or check hosts that are suspected of being infected. In some cases, non-IT staff might fulfill this duty at remote offices that do not have available IT staff. Any staff who might need to assist during major malware incidents should be designated in advance and provided with documentation and periodic training on their possible duties.

4.2.2.4 Identification Recommendations

Although active approaches typically produce the most accurate results, they are often not the fastest way of identifying infections. It might take considerable time to scan every host in an organization, and because hosts that have been disconnected or shut off will not be identified, the scan will need to be repeated. If forensic data is very recent, it might be a good source of readily available information, although the information might not be comprehensive. Manual methods are generally not feasible for comprehensive enterprise-wide identification, but they are a necessary part of identification when other methods are not sufficient. In many cases, it is most effective to use multiple approaches simultaneously or in sequence to provide the best results.

Organizations should carefully consider the possible approaches for their environment ahead of time, select a sufficiently broad range of approaches, and develop procedures and technical capabilities to perform each selected approach effectively when a malware incident occurs. Organizations should also identify which individuals or groups can assist in identification efforts. For example, identification might be performed by security administrators, system administrators, network administrators, desktop administrators, mobile device administrators, and others, depending on the sources of identification information. Organizations should ensure that everyone who might be involved in identification knows what his or her role is and how to perform necessary tasks.

4.2.3 Prioritizing Incident Response

Once a malware incident has been validated, the next activity is to prioritize its handling. NIST SP 800-61 presents general guidelines for incident prioritization; this section extends those by including additional factors to consider during prioritization.

Certain forms of malware, such as worms, tend to spread very quickly and can cause a substantial impact in minutes or hours, so they often necessitate a high-priority response. Other forms of malware, such as Trojan horses, tend to affect a single host; the response to such incidents should be based on the value of the data and services provided by the host. Organizations should establish a set of criteria that identify the appropriate level of response for various malware-related situations. The criteria should incorporate considerations such as the following:

- How the malware entered the environment and what transmission mechanisms it uses

- What type of malware it is (e.g., virus, worm, Trojan horse)

- Which types of attacker tools are placed onto the host by the malware

- What networks and hosts the malware is affecting and how it is affecting them

- How the impact of the incident is likely to increase in the following minutes, hours, and days if the incident is not contained.

4.2.4 Malware Analysis

Incident handlers can study the behavior of malware by analyzing it either actively (executing the malware) or forensically (examining the infected host for evidence of malware). Forensic approaches are safer to perform on an infected host because they can examine the host without allowing the malware to continue executing. However, sometimes it is significantly faster and easier to analyze malware by monitoring it during execution. Such active approaches are best performed on malware test systems instead of production hosts, to minimize possible damage caused by allowing the malware to execute.

Ideal active approaches involve an incident handler acquiring a malware sample from an infected host and placing the malware on an isolated test system. Test systems often have a virtualized OS image; copies of these builds can be infected, isolating any infection within the virtualized OS, and the infected image can be replaced with a known good image after the analysis is complete.[17] On such test systems, the host OS is kept uninfected so it can be used to monitor the execution of the malware within the virtualized OS. The test system should include up-to-date tools for identifying malware (e.g., antivirus software, intrusion detection systems), listing the currently running processes, and displaying network connections, as well as many other potentially helpful utilities. There are various websites and books that provide detailed

---

[17] Some malware can detect the presence of a virtualized environment and change their behavior accordingly.

instructions on setting up malware test systems and their tools; further discussion of them is outside the scope of this publication. Malware test systems are helpful not only for analyzing current malware threats without the risk of inadvertently causing additional damage to the organization, but also for training staff in malware incident handling.

Forensic approaches involve booting a forensic environment and using it to study the stored information from an infected host. The toolsets for forensic analysis greatly overlap those for active analysis; similarly, there are various websites and books available that explain how to create forensic analysis environments. There are two basic approaches: create a bootable forensic environment on write-protected removable media, or use a forensic workstation and connect it to the storage of the infected host (e.g., hard drive). The motivation for using such a trusted toolkit instead of relying on the information reported by the infected host's OS is that malware on the host may have disabled or altered the functionality of the security tools on the infected host, such as antivirus software, so that they do not report malicious activity. By running tools from a protected, verified toolkit, incident handlers can gain a more accurate understanding of the activity on the host.

4.3   Containment

Containment of malware has two major components: stopping the spread of the malware and preventing further damage to hosts. Nearly every malware incident requires containment actions. In addressing an incident, it is important for an organization to decide which methods of containment to employ initially, early in the response. Containment of isolated incidents and incidents involving noninfectious forms of malware is generally straightforward, involving such actions as disconnecting the affected hosts from networks or shutting down the hosts. For more widespread malware incidents, such as fast-spreading worms, organizations should use a strategy that contains the incident for most hosts as quickly as possible; this should limit the number of machines that are infected, the amount of damage that is done, and the amount of time that it will take to fully recover all data and services.

In containing a malware incident, it is also important to understand that stopping the spread of malware does not necessarily prevent further damage to hosts. Malware on a host might continue to exfiltrate sensitive data, replace OS files, or cause other damage. In addition, some instances of malware are designed to cause additional damage when network connectivity is lost or other containment measures are performed. For example, an infected host might run a malicious process that contacts another host periodically. If that connectivity is lost because the infected host is disconnected from the network, the malware might overwrite all the data on the host's hard drive. For these reasons, handlers should not assume that just because a host has been disconnected from the network, further damage to the host has been prevented, and in many cases, should begin eradication efforts as soon as possible to prevent more damage.

Organizations should have strategies and procedures in place for making containment-related decisions that reflect the level of risk acceptable to the organization. For example, an organization might decide that infected hosts performing critical functions should not be disconnected from networks or shut down if the likely damage to the organization from those functions being unavailable would be greater than the security risks posed by not isolating or shutting down the host. Containment strategies should support incident handlers in selecting the appropriate combination of containment methods based on the characteristics of a particular situation.

Containment methods can be divided into four basic categories: relying on user participation, performing automated detection, temporarily halting services, and blocking certain types of network connectivity. Sections 4.3.1 through 4.3.4 describe each category in detail.

4.3.1   Containment Through User Participation

At one time, user participation was a valuable part of containment efforts, particularly during large-scale incidents in non-managed environments. Users were provided with instructions on how to identify infections and what measures to take if a host was infected, such as calling the help desk, disconnecting the host from the network, or powering off the host. The instructions might also cover malware eradication, such as updating antivirus signatures and performing a host scan, or obtaining and running a specialized malware eradication utility. As hosts have increasingly become managed, user participation in containment has sharply decreased. However, having users perform containment actions is still helpful in non-managed environments and other situations in which use of fully automated containment methods (such as those described in Sections 4.3.2 through 4.3.4) is not feasible.

Effectively communicating helpful information to users in a timely manner is challenging. Although email is typically the most efficient communication mechanism, it might be unavailable during certain incidents, or users might not read the email until it is too late. Therefore, organizations should have several alternate mechanisms in place for distributing information to users, such as sending messages to all voice mailboxes within the organization, posting signs in work areas, and handing out instructions at building and office entrances. Organizations with significant numbers of users in alternate locations, such as home offices and small branch offices, should ensure that the communication mechanisms reach these users. Another important consideration is that users might need to be provided with software, such as cleanup utilities, and software updates, such as patches and updated antivirus signatures. Organizations should identify and implement multiple methods for delivering software utilities and updates to users who are expected to assist with containment.

Although user participation can be very helpful for containment, organizations should not rely on this means for containing malware incidents unless absolutely necessary. No matter how containment guidance is communicated, it is unlikely that all users will receive it and realize that it might pertain to them. In addition, some users who receive containment instructions are unlikely to follow the directions successfully because of a lack of understanding, a mistake in following the directions, or host-specific characteristics or variations in the malware that make the directions incorrect for that host. Some users also might be focused on performing their regular tasks and be unconcerned about the possible effects of malware on their hosts. Nevertheless, for large-scale incidents involving a sizable percentage of the organization's hosts in non-managed environments, user involvement in containment can significantly reduce the burden on incident handlers and technical support staff in responding to the incident.

4.3.2   Containment Through Automated Detection

Many malware incidents can be contained primarily through the use of the automated technologies described in Section 3.4 for preventing and detecting infections. These technologies include antivirus software, content filtering, and intrusion prevention software. Because antivirus software on hosts can detect and remove infections, it is often the preferred automated detection method for assisting in containment. However, as previously discussed, many of today's malware threats are novel, so antivirus software and other technologies often fail to recognize them as being malicious. Also, malware that compromises the OS may disable security controls such as antivirus software, particularly in unmanaged environments where users have greater control over their hosts. Containment through antivirus software is not as robust and effective as it used to be.

Organizations should be prepared to use other security tools to contain the malware until the antivirus signatures can perform the containment effectively, if antivirus signatures become available at all.[18] After

---

[18]   Incident handlers should also be familiar with the organization's policy and procedures for submitting copies of unknown

an organization receives updated signatures, it is prudent to test them at least minimally before deployment, to ensure that the update itself should not cause a negative impact on the organization. Another benefit of having multiple types of automated detection ready is that different detectors may be more effective in different situations. For example, detection tools that were not capable of recognizing or stopping malware when it was a new threat can sometimes be updated or reconfigured to recognize the same malware's characteristics later and stop it from spreading. Examples of automated detection methods other than antivirus software are as follows:

- **Content Filtering.** For example, email servers and clients, as well as anti-spam software, can be configured to block emails or email attachments that have certain characteristics, such as a known bad subject, sender, message text, or attachment name or type.[19] This is only helpful when the malware has static characteristics; highly customized malware usually cannot be blocked effectively using content filtering. Web content filtering and other content filtering technologies may also be of use for static malware.

- **Network-Based IPS Software.** Most IPS products allow their prevention capabilities to be enabled for specific signatures. If a network-based IPS device is inline, meaning that it is an active part of the network, and it has a signature for the malware, it should be able to identify the malware and stop it from reaching its targets. If the IPS device does not have its prevention capabilities enabled, it may be prudent during a severe incident to reconfigure or redeploy one or more IPS sensors and enable IPS so they can stop the activity. IPS technologies should be able to stop both incoming and outgoing infection attempts. Of course, the value of IPSs in malware containment depends on the availability and accuracy of a signature to identify the malware. Several IPS products allow administrators to write custom signatures based on some of the known characteristics of the malware, or to customize existing signatures. For example, an IPS may allow administrators to specify known bad email attachment names or subjects, or to specify known bad destination port numbers. In many cases, IPS administrators can have their own accurate signature in place hours before antivirus vendors have signatures available. In addition, because the IPS signature affects only network-based IPS sensors, whereas antivirus signatures generally affect all workstations and servers, it is generally less risky to rapidly deploy a new IPS signature than new antivirus signatures.

- **Executable Blacklisting.** Some operating systems, host-based IPS products, and other technologies can restrict certain executables from being run. For example, administrators can enter the names of files that should not be executed. If antivirus signatures are not yet available for a new threat, it might be possible to configure a blacklisting technology to block the execution of the files that are part of the new threat.

4.3.3 Containment Through Disabling Services

Some malware incidents necessitate more drastic and potentially disruptive measures for containment. These incidents make extensive use of a particular service. Containing such an incident quickly and effectively might be accomplished through a loss of services, such as shutting down a service used by malware, blocking a certain service at the network perimeter, or disabling portions of a service (e.g., large mailing lists). Also, a service might provide a channel for infection or for transferring data from infected hosts—for example, a botnet command and control channel using Internet Relay Chat (IRC). In either

---

malware to the organization's antivirus vendors and other security software vendors for analysis. This practice can help vendors respond more quickly to new threats. Organizations should also contact trusted parties, such as incident response organizations, when needed for guidance on handling new threats.

[19]   Generally, it is feasible only in highly managed environments to configure email clients throughout the organization to block certain emails or email attachments.

case, shutting down the affected services might be the best way to contain the infection without losing all services. This action is typically performed at the application level (e.g., disabling a service on servers) or at the network level (e.g., configuring firewalls to block IP addresses or ports associated with a service). The goal is to disable as little functionality as possible while containing the incident effectively. To support the disabling of network services, organizations should maintain lists of the services they use and the TCP and UDP ports used by each service.

From a technology standpoint, disabling a service is generally a simple process; understanding the consequences of doing so tends to be more challenging. Disabling a service that the organization relies on has an obvious negative impact on the organization's functions. Also, disabling a service might inadvertently disrupt other services that depend on it. For example, disabling email services could impair directory services that replicate information through email. Organizations should maintain a list of dependencies between major services so that incident handlers are aware of them when making containment decisions. Also, organizations might find it helpful to provide alternative services with similar functionality. For example, in a highly managed environment, if a vulnerability in an email client were being exploited by a new virus, users could be blocked temporarily from using that email client and instead directed to use a web-based email client that did not have the vulnerability. This step would help contain the incident while providing users with email access. The same strategy could be used for cases involving exploitation of vulnerabilities in web browsers and other common client applications.

4.3.4 Containment Through Disabling Connectivity

Containing incidents by placing temporary restrictions on network connectivity can be very effective. For example, if infected hosts attempt to establish connections with an external host to download rootkits, handlers should consider blocking all access to the external host (by IP address or domain name, as appropriate). Similarly, if infected hosts within the organization attempt to spread their malware, the organization might block network traffic from the hosts' IP addresses to control the situation while the infected hosts are physically located and disinfected. An alternative to blocking network access for particular IP addresses is to disconnect the infected hosts from the network, which could be accomplished by reconfiguring network devices to deny network access or physically disconnecting network cables from infected hosts.

The most drastic containment step is purposely breaking needed network connectivity for uninfected hosts. This could eliminate network access for groups of hosts, such as remote VPN users. In worst-case scenarios, isolating subnets from the primary network or the Internet might be necessary to stop the spread of malware, halt damage to hosts, and provide an opportunity to mitigate vulnerabilities. Implementing a widespread loss of connectivity to achieve containment is most likely to be acceptable to an organization in cases in which malware activity is already causing severe network disruptions or infected hosts are performing an attack against other organizations. Because a major loss of connectivity almost always affects many organizational functions, connectivity usually must be restored as soon as possible.

Organizations can design and implement their networks to make containment through loss of connectivity easier to do and less disruptive. For example, some organizations place their servers and workstations on separate subnets; during a malware incident targeting workstations, the infected workstation subnets can be isolated from the main network, and the server subnets can continue to provide functionality to external customers and internal workstation subnets that are not infected. Another network design strategy related to malware containment is the use of separate virtual local area networks (VLAN) for infected hosts. With this design, a host's security posture is checked when it wants to join the network, and also may be checked periodically while connected. The security checking is often done through network access control software by placing on each host an agent that monitors various characteristics of the host, such as OS patches and antivirus updates. When the host attempts to connect to the network, a network device such as a router requests information from the host's agent. If the host does not respond to the request or the response indicates that the host is insecure, the network device causes the host to be placed onto a separate VLAN. The same technique can be used with hosts that are already on the organization's regular networks, allowing infected hosts to be moved automatically to a separate VLAN.

Having a separate VLAN for infected hosts also helps organizations to provide antivirus signature updates and OS and application patches to the hosts while severely restricting what they can do. Without a separate VLAN, the organization might need to remove infected hosts' network access entirely, which necessitates transferring and applying updates manually to each host to contain and eradicate the malware and mitigate vulnerabilities. A variant of the separate VLAN strategy that can be effective in some situations is to place all hosts on a particular network segment in a VLAN and then move hosts to the production network as each is deemed to be clean and remediated.

4.3.5 Containment Recommendations

Containment can be performed through many methods in the four categories described above (users, automated detection, loss of services, and loss of connectivity). Because no single malware containment category or individual method is appropriate or effective in every situation, incident handlers should select a combination of containment methods that is likely to be effective in containing the current incident while limiting damage to hosts and reducing the impact that containment methods might have on other hosts. For example, shutting down all network access might be very effective at stopping the spread of malware, but it would also allow infections on hosts to continue damaging files and would disrupt many important functions of the organization.

The most drastic containment methods can be tolerated by most organizations for only a brief period of time. Accordingly, organizations should support sound containment decisions by having policies that clearly state who has authority to make major containment decisions and under what circumstances various actions (e.g., disconnecting subnets from the Internet) are appropriate.

4.4 Eradication

Although the primary goal of eradication is to remove malware from infected hosts, eradication is typically more involved than that. If an infection was successful because of a host vulnerability or other security weakness, such as an unsecured file share, then eradication includes the elimination or mitigation of that weakness, which should prevent the host from becoming reinfected or becoming infected by another instance of malware or a variant of the original threat. Eradication actions are often consolidated with containment efforts. For example, organizations might run a utility that identifies infected hosts, applies patches to remove vulnerabilities, and runs antivirus software that removes infections. Containment actions often limit eradication choices; for example, if an incident is contained by disconnecting infected hosts from the primary network, the hosts should either be connected to a separate VLAN so that they can be updated remotely, or patched and reconfigured manually. Because the hosts are disconnected from the primary network, the incident handlers will be under pressure to perform eradication actions on the hosts as quickly as possible so that the users can regain full use of their hosts.

Different situations necessitate various combinations of eradication techniques. In cases where disinfection is possible, the most common tools for eradication are antivirus software, vulnerability management technologies, network access control software, and other tools designed to remove malware and correct vulnerabilities. Automated eradication methods, such as triggering antivirus scans remotely, are much more efficient than manual methods, such as visiting infected hosts in person and running disinfection software from a CD. As described in Section 4.3.1, some situations necessitate user participation in containment and eradication activities. Providing instructions and software updates to users works in some cases, but other users might need assistance. Having formal or informal walk-up help desk areas at major facilities can also be effective and is more efficient and convenient than having IT staff locate and interrupt each affected user. During major incidents, additional IT staff members can be relieved of other duties temporarily to assist in eradication efforts. For locations without IT staff, it is often helpful to have a few people trained in basic eradication actions so that they can take care of their own hosts. Organizations should be prepared to perform a few different types of eradication efforts simultaneously if needed.

For many malware incidents, simple disinfection is not feasible, so it is necessary to rebuild all infected hosts as part of eradication efforts. Rebuilding includes the reinstallation and securing of the OS and applications (or restoration of known good OS and application backups, including the use of built-in OS rollback capabilities), and the restoration of data from known good backups. Some types of malware are extremely difficult to remove from hosts; even if they can be removed, each host's OS may be damaged, possibly to the point where the hosts cannot boot. Rebuilding is also the best eradication option when the actions performed on an infected host are unknown. If a host has multiple infections; has been infected for an extended or unknown period of time; or has had backdoors, rootkits, or other damaging attacker tools installed, other malicious actions besides the malware infections may have been performed against the host. In such cases, rebuilding the host would be the most reliable way of restoring its integrity. Also, in some cases it is faster to rebuild a host than to perform all of the analysis necessary to determine exactly what the malware has done and remove all traces of it from the host. This is particularly true in managed environments where hosts are built based on standard OS images, baselines, etc. Organizations should be prepared to rebuild hosts quickly, as needed, when malware incidents occur.

In general, organizations should rebuild any host that has any of the following incident characteristics, instead of performing typical eradication actions (disinfection):

- One or more attackers gained administrator-level access to the host.

- Unauthorized administrator-level access to the host was available to anyone through a backdoor, an unprotected share created by a worm, or other means.

- System files were replaced by a Trojan horse, backdoor, rootkit, attacker tools, or other means.

- The host is unstable or does not function properly after the malware has been eradicated by antivirus software or other programs or techniques. This indicates that either the malware has not been eradicated completely or that it has caused damage to important system or application files or settings.

- There is doubt about the nature of and extent of the infection or any unauthorized access gained because of the infection.

If a malware incident does not have any of these characteristics, then it is typically sufficient to eradicate the malware from the host instead of rebuilding the host.

Eradication can be frustrating because of the number of hosts to clean up and the tendency to have additional infections and reinfections occurring for days, weeks, or months.[20] Incident handlers should periodically perform identification activities to identify hosts that are still infected and estimate the

---

[20]   Instances of a particular type of malware might reside within an organization indefinitely, regardless of eradication efforts. For example, malware might be captured in host backups; restoration of a backup could also restore the malware. Also, malware might infect removable media that then sits unused for an extended period of time. Years after the initial infection, the removable media could be accessed, and the malware could attempt to infect the host.

success of the eradication. A reduction in the number of infected hosts would demonstrate that the incident response team was making progress and would help the team choose the best strategy for handling the remaining hosts and allocate sufficient time and resources. It can be tempting to declare an incident resolved once the number of infected hosts has dropped significantly from the original numbers, but the organization should strive to reduce the suspected numbers of infected and vulnerable machines to low enough levels that if they were all connected to the network at once and the vulnerable machines all became infected, the overall impact of the infections would be minimal.

4.5 Recovery

The two main aspects of recovery from malware incidents are restoring the functionality and data of infected hosts and removing temporary containment measures. Additional actions to restore hosts are not necessary for most malware incidents that cause limited host damage (for example, an infection that simply altered a few data files and was completely removable with antivirus software). As discussed in Section 4.4, for malware incidents that are far more damaging, such as Trojan horses, rootkits, or backdoors, corrupting thousands of system and data files, or wiping out hard drives, it is often best to first rebuild the host, then secure the host so that it is no longer vulnerable to the malware threat. Organizations should carefully consider possible worst-case scenarios, such as a new malware threat that necessitates rebuilding a large percentage of the organization's workstations, and determine how the hosts would be recovered in these cases. This should include identifying who would perform the recovery tasks, estimating how many hours of labor would be needed and how much calendar time would elapse, and determining how the recovery efforts should be prioritized.

Determining when to remove temporary containment measures, such as suspended services (e.g., email) or connectivity (e.g., Internet access, VPN for telecommuters), is often a difficult decision during major malware incidents. For example, suppose that email has been shut down to stop the spread of a malware infection while vulnerable hosts are patched and infected hosts undergo individual malware containment, eradication, and recovery measures. It might take days or weeks for all vulnerable hosts to be located and patched and for all infected hosts to be cleaned, but email cannot remain suspended for that period of time. When email service is restored, it is almost certain that an infected host will begin spreading the malware again at some time. However, if nearly all hosts have been patched and cleaned, the impact of a new malware infection should be minimal. Incident response teams should strive to keep containment measures in place until the estimated number of infected hosts and hosts vulnerable to infection is sufficiently low that subsequent incidents should be of little consequence. Incident handlers should also consider alternative containment measures that could adequately maintain containment of the incident while causing less of an impact on the normal functions of the organization. However, even though the incident response team should assess the risks of restoring the service, management should ultimately be responsible for determining what should be done, based on the incident response team's recommendations and management's understanding of the business impact of maintaining the containment measures.

4.6 Lessons Learned

When a major malware incident occurs, the primary individuals performing the response usually work intensively for days or weeks. As the major handling efforts end, the key people are usually mentally and physically fatigued, and are behind in performing other tasks that were pending during the incident handling period. Consequently, the lessons learned phase of incident response might be significantly delayed or skipped altogether for major malware incidents. However, because major malware incidents can be extremely expensive to handle, it is particularly important for organizations to conduct robust lessons learned activities for major malware incidents. Although it is reasonable to give handlers and other key people a few days to catch up on other tasks, review meetings and other efforts should occur expeditiously, while the incident is still fresh in everyone's minds. The lessons learned process for malware incidents is no different than for any other type of incident. Examples of possible outcomes of lessons learned activities for malware incidents are as follows:

■ **Security Policy Changes.** Security policies might be modified to prevent similar incidents. For example, if connecting personally owned mobile devices to organization laptops caused a serious infection, modifying the organization's policies to secure, restrict, or prohibit such device connections might be advisable.

■ **Awareness Program Changes.** Security awareness training for users might be changed to reduce the number of infections or to improve users' actions in reporting incidents and assisting with handling incidents on their own hosts.

■ **Software Reconfiguration.** OS or application settings might need to be changed to support security policy changes or to achieve compliance with existing policy.

■ **Malware Detection Software Deployment.** If hosts were infected through a transmission mechanism that was unprotected by antivirus software or other malware detection tools, an incident might provide sufficient justification to purchase and deploy additional software.

■ **Malware Detection Software Reconfiguration.** Detection software might need to be reconfigured in various ways, such as the following:

– Increasing the frequency of software and signature updates

– Improving the accuracy of detection (e.g., fewer false positives, fewer false negatives)

– Increasing the scope of monitoring (e.g., monitoring additional transmission mechanisms, monitoring additional files or file systems)

– Changing the action automatically performed in response to detected malware

– Improving the efficiency of update distribution.

Appendix A—Glossary

Selected terms used in the guide are defined below.

Antivirus Software: A program that monitors a computer or network to identify all major types of malware and prevent or contain malware incidents.

Backdoor: A malicious program that listens for commands on a certain Transmission Control Protocol (TCP) or User Datagram Protocol (UDP) port.

Disinfecting: Removing malware from within a file.

False Negative: An instance in which a security tool intended to detect a particular threat fails to do so.

False Positive: An instance in which a security tool incorrectly classifies benign content as malicious.

Malware: A program that is covertly inserted into another program with the intent to destroy data, run destructive or intrusive programs, or otherwise compromise the confidentiality, integrity, or availability of the victim's data, applications, or operating system.

Mobile Code: Software that is transmitted from a remote host to be executed on a local host, typically without the user's explicit instruction.

On-Access Scanning: Configuring a security tool to perform real-time scans of each file for malware as the file is downloaded, opened, or executed.

On-Demand Scanning: Allowing users to launch security tool scans for malware on a computer as desired.

Phishing: Tricking individuals into disclosing sensitive personal information through deceptive computer-based means.

Quarantining: Storing files containing malware in isolation for future disinfection or examination.

Rootkit: A collection of files that is installed on a host to alter the standard functionality of the host in a malicious and stealthy way.

Signature: A set of characteristics of known malware instances that can be used to identify known malware and some new variants of known malware.

Appendix B—Acronyms and Abbreviations

Selected acronyms and abbreviations used in the guide are defined below.

ACL	Access Control List
CSRC	Computer Security Resource Center
DDoS	Distributed Denial of Service
DMARC	Domain-based Message Authentication, Reporting & Conformance
DNS	Domain Name System
FAQ	Frequently Asked Questions
FISMA	Federal Information Security Management Act
HTML	Hypertext Markup Language
HTTP	Hypertext Transfer Protocol
ICMP	Internet Control Message Protocol
ID	Identification
IDS	Intrusion Detection System
IETF	Internet Engineering Task Force
IP	Internet Protocol
IPS	Intrusion Prevention System
IT	Information Technology
ITL	Information Technology Laboratory
MAC	Media Access Control
NAP	Network Access Protection
NAT	Network Address Translation
NIST	National Institute of Standards and Technology
NSRL	National Software Reference Library
OMB	Office of Management and Budget
OS	Operating System
PIN	Personal Identification Number
RFC	Request for Comment
SIEM	Security Information and Event Management
SMTP	Simple Mail Transfer Protocol
SP	Special Publication
TCP	Transmission Control Protocol
UDP	User Datagram Protocol
USB	Universal Serial Bus
US-CERT	United States Computer Emergency Readiness Team
VBScript	Visual Basic Script
VLAN	Virtual Local Area Network
VPN	Virtual Private Network

Appendix C—Resources

The following lists provide examples of resources that may be helpful in understanding malware and in preventing and handling malware incidents.

Organizations

Organization	URL
Anti-Phishing Working Group (APWG)	http://www.antiphishing.org/
Anti-Virus Information Exchange Network (AVIEN)	http://www.avien.org/
Computer Antivirus Research Organization (CARO)	http://www.caro.org/
Cooperative Association for Internet Data Analysis (CAIDA)	http://www.caida.org/
European Institute for Computer Antivirus Research (EICAR)	http://www.eicar.org/
Internet Storm Center (ISC)	http://isc.incidents.org/
Securelist	http://www.securelist.com/en/
United States Computer Emergency Readiness Team (US-CERT)	http://www.us-cert.gov/
Virus Bulletin	http://www.virusbtn.com/
WildList Organization International	http://www.wildlist.org/

Other Technical Resource Documents

Resource Name	URL
FTC, *How Not to Get Hooked by a "Phishing" Scam*	http://ftc.gov/bcp/conline/pubs/alerts/phishingalrt.htm
IETF, RFC 2267, *Network Ingress Filtering: Defeating Denial of Service Attacks Which Employ IP Source Address Spoofing*	http://www.ietf.org/rfc/rfc2267.txt
NIST, SP 800-28 Version 2, *Guidelines on Active Content and Mobile Code*	http://csrc.nist.gov/publications/PubsSPs.html#800-28
NIST, SP 800-37 Revision 1, *Guide for Applying the Risk Management Framework to Federal Information Systems*	http://csrc.nist.gov/publications/PubsSPs.html#800-37
NIST, SP 800-40 Revision 3, *Guide to Enterprise Patch Management Technologies*	http://csrc.nist.gov/publications/PubsSPs.html#800-40-rev3
NIST, SP 800-41 Revision 1, *Guidelines on Firewalls and Firewall Policy*	http://csrc.nist.gov/publications/PubsSPs.html#800-41
NIST, SP 800-45 Version 2, *Guidelines on Electronic Mail Security*	http://csrc.nist.gov/publications/PubsSPs.html#800-45
NIST, SP 800-53 Revision 3, *Recommended Security Controls for Federal Information Systems and Organizations*	http://csrc.nist.gov/publications/PubsSPs.html#800-53
NIST, SP 800-61 Revision 2, *Computer Security Incident Handling Guide*	http://csrc.nist.gov/publications/PubsSPs.html#800-61-rev2
NIST, SP 800-70 Revision 2, *Security Configuration Checklists Program for IT Products*	http://csrc.nist.gov/checklists/
NIST, SP 800-86, *Guide to Applying Forensic Techniques to Incident Response*	http://csrc.nist.gov/publications/PubsSPs.html#800-86
NIST, SP 800-92, *Guide to Computer Security Log Management*	http://csrc.nist.gov/publications/PubsSPs.html#800-92

Resource Name	URL
NIST, SP 800-94, *Guide to Intrusion Detection and Prevention Systems (IDPS)*	http://csrc.nist.gov/publications/PubsSPs.html#800-94
NIST, SP 800-115, *Technical Guide to Information Security Testing and Assessment*	http://csrc.nist.gov/publications/PubsSPs.html#800-115
NIST, SP 800-117, *Guide to Adopting and Using the Security Content Automation Protocol (SCAP)*	http://csrc.nist.gov/publications/PubsSPs.html#800-117
NIST, SP 800-128, *Guide for Security-Focused Configuration Management of Information Systems*	http://csrc.nist.gov/publications/PubsSPs.html#800-128
NIST, SP 800-147. *BIOS Protection Guidelines*	http://csrc.nist.gov/publications/PubsSPs.html#800-147

www.ingramcontent.com/pod-product-compliance
Lightning Source LLC
Chambersburg PA
CBHW081358170526
45166CB00010B/3126